Wood

by Melanie Mitchell

first step nonfiction

Lerner Publications · Minneapolis

Look at all of the **wood**.

There are many kinds and colors of wood.

Wood comes from trees.

Trees are cut into pieces at a **sawmill**.

Pieces of wood can be used to build **houses**.

Some houses are all wood.

Tables are made of wood.

Chairs are made of wood.

Many toys are made of wood.

This rocking horse is made of wood.

Baseball players use bats made of wood.

People write with pencils made of wood.

Wood is burned to make **heat**.

You can cook food over a **campfire**.

Wood is used for many things.

Wood is important to us.

How Old Are Trees?

Trees don't just grow taller every year. Tree trunks get wider, too. Every year a tree adds a new layer of wood under its bark. These new layers form rings inside the tree trunk. When a tree is cut down, you can see the rings. A thick ring means the tree grew a lot that year. A thin ring means it didn't grow as much due to bad weather or other problems. By counting the rings, you can get an idea of how old a tree is.

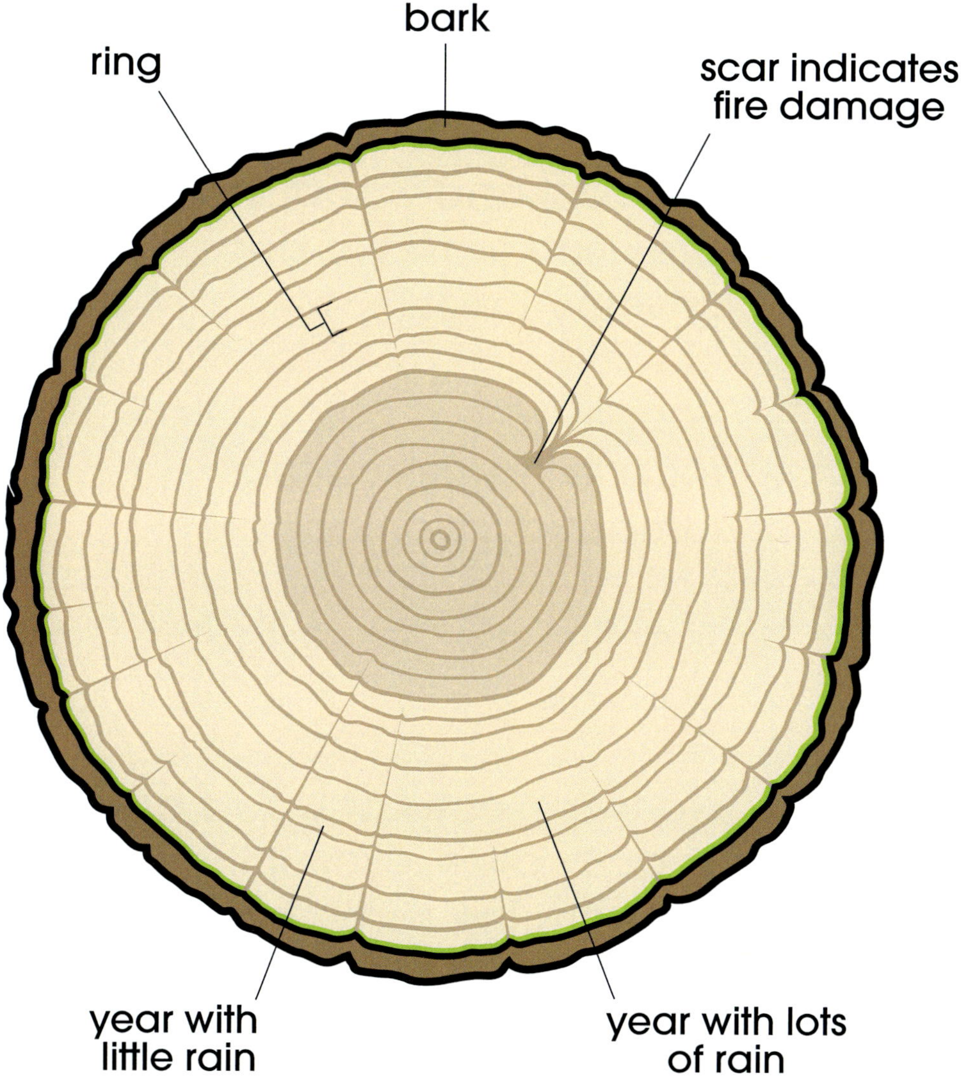
bark
ring
scar indicates fire damage
year with little rain
year with lots of rain

Wood Facts

- Sawdust and wood shavings are used to make paper grocery bags.

- People who cut down trees for a living are called lumberjacks or loggers.

- Termites are insects that eat wood.

- No two pieces of wood are exactly alike, even if they are cut from the same tree.

- A cord of wood is wood that is stacked 4 feet high, 4 feet wide, and 8 feet deep. One cord of wood can make 7.5 million toothpicks.

- The state of Maine produces more toothpicks than any other state.

- Petrified wood is wood from millions of years ago that has turned into stone.

Glossary

campfire – an outdoor fire used for cooking and heat

heat – warmth provided for people or places

houses – buildings people live in

sawmill – a place where wood is cut into pieces

wood – hard material under the bark of trees

Index

The photographs in this book are reproduced through the courtesy of: © PhotoDisc Royalty Free Images, front cover, pp. 2, 6; © Michael Gabridge/Visuals Unlimited, pp. 3, 22 (bottom); © Esbin-Anderson/Photo Network, p. 4; © Tom Tracy/Photo Network, pp. 5, 22 (second from bottom); © Tom Bean/CORBIS, pp. 7, 22 (middle); © Wm. G. Johnson/Visuals Unlimited, p. 8; © Jeff Greenberg/Visuals Unlimited, pp. 9, 15, 22 (top); © Inga Spence/Visuals Unlimited, p. 10; © Jacqui Hurst/CORBIS, p. 11; © Tom Uhlman/Visuals Unlimited, p. 12; © Michael Astalos/Photo Network, p. 13; © Cynthia Salter/Photo Network, pp. 14, 22 (second from top); © Diane Meyer, p. 16; © Mark Newman/ Photo Network, p. 17.

Illustration p. 19 by Laura Westlund.

Lerner Publications Company
A division of Lerner Publishing Group, Inc.
241 First Avenue North
Minneapolis, MN 55401 USA

For reading levels and more information, look up this title at www.lernerbooks.com.

Library of Congress Cataloging-in-Publication Data

Mitchell, Melanie S.
Wood / by Melanie Mitchell.
p. cm. — (First step nonfiction)
Summary: An introduction to wood and its uses in everyday life.
ISBN: 978–0–8225–4618–4 (lib. bdg. : alk. paper)
1. Wood—Juvenile literature. [1. Wood.] I. Title. II. Series.
TA419 .M56 2003
620.1'2—dc21 2002006472

Manufactured in the United States of America
2 - 39103 - 4651 - 4/19/2016